WRITTEN BY EMMANUEL THREATT
ILLUSTRATED BY BRIAN BLACKMORE

Copyright (C) 2020 by Kardboard Kids, LLC

All rights reserved. No part of this book may be reproduced or used in any manner without written permission of the copyright owner except for the use of quotations in a book review.

KARDBOARDKIDS.COM
@KARDBOARDKIDS

Red, blue and green lights flash all around.

Switches, knobs and buttons click and clack as a low hum sounds.

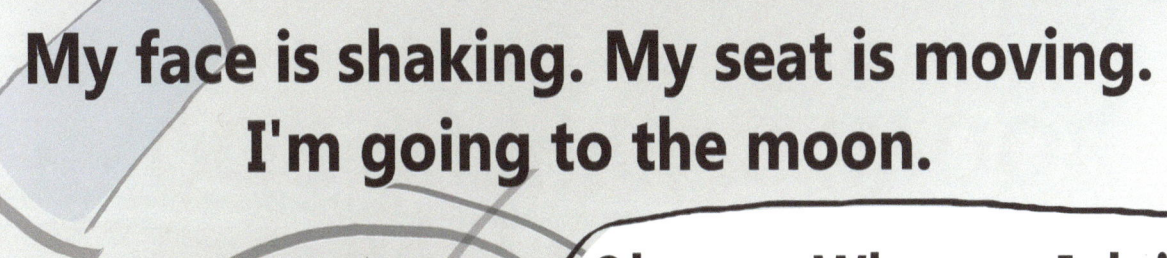

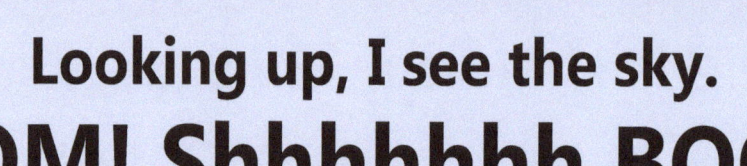

Looking up, I see the sky.
BOOM! Shhhhhhh BOOM!
I'm about to fly!

Thump, Thump... Thump, Thump.
I hear my heart. I hear it pound.
I'm feeling nervous,
but I'm not off the ground.

Will I need a watch
to get back before dinner?
Will I need a coat?
On the moon is it winter?

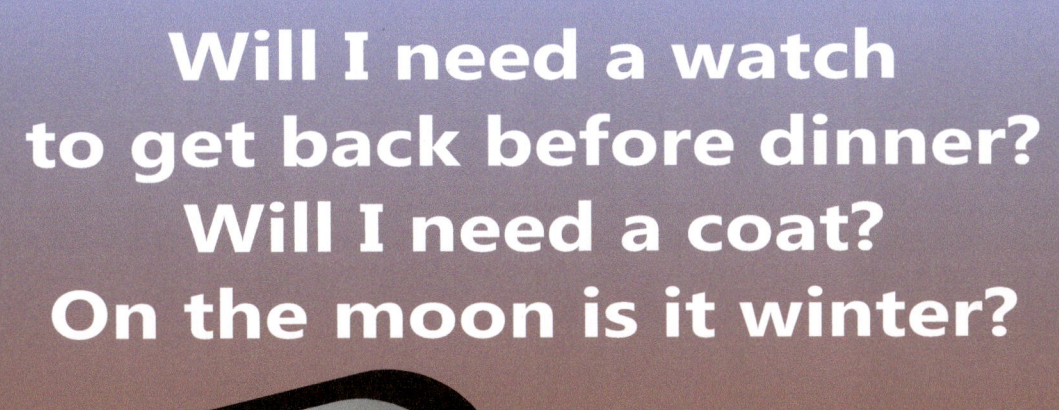

The thoughts in my mind sound REALLY, REALLY LOUD, but the rocket ship has blasted way WAY off the ground!

Through the clouds,
toward the stars...
The day turns to night.
Above the world,
I zoom past it all
and see a satellite.

I see the moon get closer and closer.
Up and down, left and right.
This feels like a rollercoaster.

Milton Keynes UK
Ingram Content Group UK Ltd.
UKHW050956161023
430695UK00002B/8